I0463252

SOVIET SCIENCE: THIRTY YEARS

S. I. Vavilov

Soviet Science: Thirty Years
by S. I. Vavilov

Published by Prism Key Press, 2010.
Website: www.prismkeypress.com

ISBN 1453766049

Cover by Claude Kipper

The three decades that have passed since the October days of 1917 have brought about, on the territory of the onetime Russian empire, such social and economic change, such historical developments, as to reshape the very foundations of life in the country. Never before has human history, has the development of society, witnessed such momentous revolutionary upheavals as this transformation of old Russia into a classless, socialist stale based on the broad democracy of the Stalin Constitution into a close-knit community of Soviet peoples, with a heroic army that has attained unexampled victory, with a huge new industry and an agriculture of an entirely new type.

The most far-reaching conclusions and forecasts of the teaching of Marx, Engels, Lenin and Stalin on the development of society have begun to be realized in the land of Soviets. For the first time in human history, scientific theory guides the building of a new slate.

And this mighty tide of history has carried with it, irresistibly, all science as a whole. Thirty Soviet years have effected a complete metamorphosis of science, both in scope and in nature. Of the scientific traditions of old Russia, only that which was progressive has gone on into the new life.

The extent and the substance of the changes that have taken place will be more easily grasped after a brief glance at the past, at the foots from which the new conditions have produced the Soviet science we Know today.

Science, in content, form, and purpose, is fundamentally social, collective. It is invariably, in its every branch, the sum of knowledge attained by many different people, by past generations and by contemporaries. It is the composite product of collective labors. The facts and conclusions which it comprises are expressed in the form of concepts, definitions, and formulae; they are recorded in writing or in print. The purpose of all this is to facilitate the communication of knowledge to other people, to one's class, one's stale, to humanity as a whole. Finally, and this is most important, science is a powerful instrument helping to disclose new productive forces in nature and new means of production. It gives man the means of struggle and of defense. Therefore, science comes into being and develops simultaneously with the rise and development of society, as an inevitable consequence and at the same time an indispensable condition for this development.

In Russia, the development of science began many centuries ago. Between the tenth and the twelfth centuries, it appears to have maintained the same level as in other European countries. For this we have the evidence both of writings of that period and of. material relics, particularly architectural. The invasion of the Tatars and Mongols, however, interrupted the normal growth of science in Russia. Progress was retarded for .several centuries after. The rise of secular schools was hampered, and the science of the churches and monasteries pursued aims' that had nothing in common with the progressive tendencies of natural science and technology. Clerical science was fettered and weighed down by Byzantine inertia and conservatism, by the "spiritual dictatorship of the church," as Engels puts it. Only in the seventeenth century did secular science begin to assert itself in Russia. One of its early expressions was the attempt of Boris Godunov to found a university in Moscow — a plan realized, somewhat later, in the founding of the "Slavo-Greco-Latin Academy," Moscow's first institution of higher learning. Initially, of course, this institution was concerned only with teaching, and not with scientific research.

Science began to advance rapidly during the reign of Peter I, when the interests of the state called for a considerable expansion and consolidation of industry, commerce, and the art of warfare. Feudal Russia was a backward state, both economically and culturally, as compared with Western Europe. This was due, in considerable measure, to political causes of an extraneous nature. It should be remembered that the effects of the Tatar and Mongol incursions were not entirely wiped out in Russia until the latter part of the eighteenth century. In the meanwhile, during the sixteenth and seventeenth centuries, Western Europe, entering the capitalist phase of its history, had witnessed the growth of a new and remarkable natural science the science of such men as Copernicus, Galileo, Kepler, Descartes, and Newton. Under Peter, however, this science, so new both in content and in style, took root in Russia with amazing rapidity. In the middle of the eighteenth century the St. Petersburg Academy of Sciences, which Peter had founded in 1725, was the scene of the scientific labors of M. V. Lomonosov, a man whose genius and achievements have been really grasped and appraised only in recent times, after a lapse of some two centuries. Lomonosov's work and attainments in the fields of physics, chemistry, astronomy, construction of instruments, geology, geography, language, and history would have done (honor to a whole academy, not to speak of one man. Pushkin called him "Russia's' university." He was the Russian people's swift response to the new opportunities for scientific development which had appeared, at long last, in the reign of Peter I.

Peter's foresight in basing the new Russian science on a central academy was completely justified in the course of the eighteenth century. The new Academy soon began to render useful service to the state in questions of technology and in the study of the country's geography, population, and natural resources. It set vigorously and successfully to work on the innumerable problems that had accumulated: questions of the Russian people's history and ethnography, of Russian grammar, of the country's climate. The St. Petersburg Academy members zealously promoted high school and university training for the youth. The Academy engaged in

7

publication to an extent amazing for that period, bringing Russian society at large its first knowledge of the finest classics of science and literature. Members of the Academy initiated the organization of new scientific institutions, universities, and associations of scientists. In 1755, a university, proposed and planned by M. V. Lomonosov, was founded in Moscow. It was not long before this university became an important and independent scientific center.

Unquestionably, in the period between its foundation and the end of the eighteenth century, the St. Petersburg Academy of Sciences contributed fundamentally to both Russian and world science. Here, on the banks of the Neva, native Russian scientists worked in close cooperation with foreign scientists, as for example Euler and Pallas, over the most important problems of science in that period. Here a strong foundation was built up for the atomic theory. It was here that the law of conservation of matter in chemical reactions was first proved experimentally, by M. V. Lomonosov. It was here that V. V. Petrov conducted his experiments directed against the phlogiston theory, and that physical chemistry took shape as a separate science. It was in St. Petersburg that Lomonosov established the existence of an atmosphere around the planet Venus. A wealth of important material concerning Russian flora, fauna, geography, and ethnography was compiled by S. P. Krashemrmikov, I. I. Lepekhin, N. Y. Ozeretsikovsky, V. M. Severgin, P. S. Pallas, and S. G. Gmelin. Of great significance were the investigations into Russian history conducted by V. N. Tatishchev, M. V. Lomonosov, and F. I. Miller. The profundity and importance of V. K. Tredyakovsky's philological studies are only now beginning to be appreciated.

Peter's successors on the Russian throne did not share in his respect for science, did not realize its importance to the state. At best, they tolerated the Academy of Sciences as an appendage necessary for the adornment of a European court. The Academy, the universities, the scientific associations received very little real assistance, either moral or material, from the state. Men of science were left to their own resources. There was no longer that tie between science and the life of the state which Peter had had in mind in setting up the Academy.

This, at best negligent and contemptuous, attitude of the tsarist government towards the problems of science became traditional, persisting until the very eve of the October Revolution. Purely by inertia, certain appropriations for scientific work continued to figure in the state budget; but their amount would remain unchanged for decades on end. The new scientific centers taking shape in Kazan, Kharkov, Kiev, and other provincial towns had a difficult and uphill struggle to overcome, now the stubborn resistance, now the complete indifference, of the government. Scientific institutions were regarded principally as a supply center for specialists, professors, teachers, engineers — categories that a modern European state could not very well get along without, whether it liked the idea or not. Research, creative scientific work, inventions, as a rule were deigned no notice, and even at best received but slight support and encouragement. But the Russian people had always been eager for knowledge, and what they had glimpsed of the nature and prospects of modern science intensified this urge. Self-taught inventors appeared. There was the famous Ivan Kulibin, of Nizhni Novgorod, and there was many another who unfortunately did not rise to fame, for lack of timely support. The provincial university newly established in distant Kazan fostered, the genius of that great Russian mathematician, N. I. Lobachevsky, who is often deservedly called "the Copernicus of geometry." Let the reader pause for a moment to realize how far Kazan was then removed from all cultural centers, how backward and isolated. Only then will he fully understand how remarkable it was that such a spot should produce Lobachevsky's subtle and penetrating mathematical concepts, concepts that for decades remained above the understanding of the world's greatest mathematicians. Some time later, this same Kazan produced and developed the splendid Russian school of chemistry which gave the world such men as N. N. Zinin, discoverer of aniline; A. M. Butlerov, one of the founders of modern organic chemistry; V. V. Markovnikov, and A. M. Zaitsev.

The class composition of the men of science in pre-revolutionary Russia was distinguished by the following important feature :

It was chiefly the "lower classes" — children of peasants, commoners, petty officials — that went in for scientific work with eager interest. So it was at the beginning of the eighteenth century, and so it continued for some two hundred years. Lomonosov was not the only man of science sprung from peasant stock. Few members of the ruling classes — the wealthy nobility and the bourgeoisie — allowed their children to devote themselves to learning. It was not a paying profession—Its prospects were hazy, and it involved hard work. Again, there were many who regarded science, not without foundation, as an ideological threat to their class rule. With the rise of the revolutionary movement in Russia, and the accentuation of class antagonisms, this feature of the composition of Russian scientific circles strongly influenced the development of Russian science, giving it a democratic trend. There was a comparatively narrow group of ''official" scientists, which rendered loyal service to the forces of reaction and did its best to strangle every hint of progress and innovation in science; but the Russian scientists as a Whole were in a state of constant opposition — timid and covert, it is true — to the tsarist government, which failed to realize the importance of science and the prospects before it.

Towards the end of the eighteenth century, besides the St.

Petersburg Academy, as the official, court representative of science, increasing importance began to' attach to scientific beginnings in other parts of the country, and particularly in Moscow. When Moscow University celebrated its centenary, in 1855, its list of staff professors for the hundred-year period comprised 254 names, many those of outstanding scientists in the different fields. The theory and history of literature: A. F. Merzlyakov, poet and scientist, himself a former student of the university, and Academy members S, P. Shevyryov and F. I. Buslayev. World history: Academy member M. P. Pogodin and Professor T. N. Granovsky. Russian history: among others, the famous Professor S. M. Solovyov. Physics and mathematics: the well-known astronomer D. M. Perevoshchikov; the mathematician and physicist N. D. Brashman; the gifted physicist, philosopher, and agricultural expert M. G. Pavlov; the eminent physicist and meteorologist M. F. Spassky. Biology: the zoologist K. F. Rulye. Thus, science in Moscow was growing and developing in every field.

The nineteenth century, age of development of capitalism, of steam and electricity, brought a new advance of science and technology in Western Europe, which, in the latter part of the century, spread also to America and Japan. Russia, too, experienced a rising tide of scientific activity. Splendid new men came to the fore in all the country's scientific centers — in the Academy, the universities and the specialized institutions of higher learning. The .work of N. I. Lobachevsky in the field of geometry, and of M. V. Ostrogradsky, Sophia Kovalevskaya, and P. L. Chebyshev in mathematical analysis, sent the fame of Russian mathematics ringing round the world. Many remarkable discoveries were made in the field of technical physics. The voltaic arc was produced, for the first time in history, by L. Y. Kraft and V. V. Petrov. Academy member B. S. Jaeobi discovered and

11

developed the technique of galvano-plastics, and constructed an original telegraph and the first motorboat, besides many other important practical discoveries. It was in Russia that the first practical sources of electric light came into being: P. N. Yablochkov's arc candle, and A. N. Lodygin's incandescent lamp, the first of its kind. JRadio was first discovered by the Russian A. S. Popov. Academy member and St. Petersburg University professor E. G. Lenz was one of the founders of classical electromagnetism (Lenz's law and rules). The greatest discovery of the nineteenth century in the field of chemistry — the periodic table of chemical elements — was made in St. Petersburg, by D. I. Mendeleyev. The Pulkovo Observatory, built towards the middle of the century, was for several decades the "astronomical capital of the world." Fundamental discoveries in the fields of embryology, microbiology, and physiology are bound up with the names of the great Russian biologists — K. Baer, A. O. Kovalevsky, I. I. Mechnikov, S. N. Vinogradskj, I. M. Sechenov, I. P. Pavlov. Geographical, ethnographical and archeological discoveries of great importance were made by N. M. Przhevalsky, N. N. Miklukho-Maklay, P. A. Kropotkin, P. K. Kozlov, and others. Nineteenth-century Russian science made basic contributions in the fields of orientalogy, language, and Russian and world history. II is impossible in so brief an article even to list all the outstanding scientific achievements attained by Russian scientists in various fields in the course of the nineteenth century.

Surely, eloquent testimony to the Russian people's urge for knowledge, to their talent and ability !

In spite of all this, however, Russian science in the nineteenth century did not become the powerful force it might have been, did not develop into a comprehensive, consistent, and systematic movement. It was not sufficiently bound up with life, and failed to produce what might be called, in chemical terms, a "chain" process of development of science and technology in

Russia. It was no more, than the mechanical sum of the activities of individual outstanding scientists. Only in rare cases were scientists able to found schools, to find assistants and disciples who would carry on their work. Many an important labor begun by a Russian scientist ceased with its author's death and was consigned to oblivion. Sometimes, such works were continued — abroad. This situation was caused, first and foremost, by the tsarist government's failure to appreciate Russian science, by the contempt and suspicion in which it held native endeavor and prospects. As the need for science and technology developed, the government preferred to import them, ready-made, from abroad; and Russian scientific labors, as a rule, received none but the most insignificant material assistance from the government. Scientific research, as a profession, was limited to a very small number of persons who were retained in the universities for post-graduate work, budget appropriations for which were less than negligible. Most of the young people graduating from colleges and universities entered the fields of school teaching or industry, or engaged in other practical activities very far removed from science. Thus. Russia in the nineteenth century had many brilliant scientists and could pride herself upon a lengthy roll of momentous discoveries and inventions; but, with only rare exceptions, she had no systematically developing national science.

This discrepancy between the people's latent abilities, their aspiration to knowledge, on the one hand, and the lack of government support, on the other, became particularly marked during the last few decades before the revolution. In this period we note numerous and broadly conceived attempts to create, besides the official science vegetating in state institutions, a science supported by the public, independent of government subsidies. Numerous private colleges appeared in St. Petersburg, Moscow, and other cities. Particularly successful in St. Petersburg were the

Bestuzhev courses, and the Psycho-Neurological Institute founded by Professor Lesgaft. In Moscow, a private university for women (the Women's Higher Courses) was founded and developed rapidly. There were also the Gerye courses, and the Golitsyn agricultural courses. Each of these schools offered a systematic higher education in one or several fields. Again, besides such higher schools of the generally accepted type, institutions known as "people's universities" began to spring up, almost spontaneously, in a number of cities, particularly Moscow. The people's universities engaged in the organization of popular lectures, in cycles covering the various fields of learning and on individual scientific problems. Delivered by prominent men of science and university professors, these lectures proved a great success, and regularly drew large audiences, not only of intellectuals, but also of advanced workers. Some of the people's universities also organized laboratory work, and even trips through Russia and abroad devoted to the study of botany, geology, archaeology, and the arts. In 1911, on funds contributed by A. L. Shanyavsky, a large building was erected on the Miuss Square, Moscow, to be used as a popular university. The Shanyavsky university had well-equipped lecture halls and laboratories, and a very good library. Active centers of interesting and useful work were the scientific societies — for example, the Society of Amateurs in the Natural Sciences, Anthropology, and Ethnography, and the Natural History Society, both in Moscow. The first-named society founded an institution, the Polytechnical Museum, which to this day promotes the dissemination of scientific and technological knowledge among the population of Moscow. A society of this type would comprise a number of groups, devoted to various ramifications of its chosen field of work. At their meetings, these groups engaged in lively discussions of the latest works of Moscow scientists. Many, of the city's most prominent men of science were

14

active participants in these societies, among them the renowned botanist K. A. Timiryazev, the eminent physicist P. N. Lebedev, and the founder of Russian aeronautics, N. E. Zhukovsky. Important work was done in St. Petersburg by the Russian Physical and Chemical Society, which united the labors of all the physicists and chemists of the period. There were also active scientific societies in Kazan, Kharkov, Nizhni Novgorod and a number of other cities. An even more vivid indication of public interest in science were the big St. Petersburg and Moscow congresses of naturalists and physicians. In all, there were twelve such congresses, of which the last, convened in Moscow at the very end of 1909, was particularly indicative. This congress had an attendance of something like six thousand. In other words, it attracted almost all the country's scientific forces in the field of medicine and natural history, down to university students of the graduating classes. Its general meetings heard reports on the most debated scientific topics of the day: on high nervous activity, by Academy member I. P. Pavlov, and on the theory of relativity, by N. A. Umov, professor of physics at Moscow University. At a session of the physics section, P. N. Lebedev reported on his amazingly delicate and important experiments in the field of light pressure. The twelfth congress of naturalists and physicians was the last and most impressive demonstration of the vigour and the quality of Russian science before the revolution.

A year after this congress, however, Moscow was the scene of events which reflected all too clearly the tragic situation of science in tsarist Russia. Student unrest at Moscow University, which was brought to the surface' towards the end of 1910 in connection with the death of Tolstoy, was seized upon by the Moscow police as a pretext for invading the university. Police officers, and even the Moscow Chief of Police himself, began to appear in the professors' stands in the lecture halls where the

15

students held their meetings. The rectors (Professor A. A. Manuilov, Academy member V. I. Vernadsky, and Professor P. A. Minakov) and most of the progressively-minded professors handed in their resignations, which were at once accepted by the Ministry of Public Education, then headed by Professor Kasso. Thus, for many years to come — in fact, until the revolution — Moscow University was deprived of the very core of its teaching staff. Outstanding scientists were replaced by the first chance applicants. Scientific activity declined almost to nil. In order to train young scientists, the Ministry of Public Education was obliged to select politically un-suspect groups from among the graduating students and send them abroad to study, as in the days of Peter I.

Symptomatic of the period between the revolutions of 1905 and October 1917 was the immediate public support accorded to the scientists resigning from Moscow University. Many of these men continued teaching, and resumed their research work, in the private colleges mentioned above, or in the Shanyavsky people's university. Even the Imperial Academy of Sciences, or rather, the more liberally inclined of its members, made an effort to assist them. Considerable funds were collected for the building of research institutes, two of which (the Institute of Physics, on Miuss Square, and the Institute of Experimental Biology, on Vorontsovo Polye) were actually completed just before the October Revolution. Among the men who left Moscow University was the brilliant Russian physicist, and experimenter P. N. Lebedev, who has already been mentioned above. Lebedev continued his work in a basement apartment in one of Moscow's side streets (at No. 20, Myortvy Lane). This basement was the scene of his last experiments, an interesting research into the' nature of terrestrial magnetism. On March 14, 1912, Lebedev died. He was only 46, and it is hardly to be doubted that his death was hastened by the tragedy of Moscow University.

Publicly supported scientific activities continued to develop, despite the obstacles which the government put in their way, and during the first world war rendered considerable assistance to the front. Such public organizations as the Zemstvo Union and the Union of Towns sponsored scientific work on the development of gas masks and organized the manufacture of X-ray apparatus, telephones, thermometers, etc In Petrograd, the Academy of Sciences promoted public effort by organizing a large committee for the study of Russia's natural productive forces. The numerous sections of this committee engaged in the study of problems of technical physics, geology, and chemistry, the solution of which was of great assistance to the country's war industries.

And this wartime public scientific effort subsequently became a factor in many respects facilitating the accomplishment of the tremendous tasks which confronted science immediately after the victory of the socialist revolution.

The qualitative level of Russian science in the last decade preceding the revolution was very high. Brilliant work was being done in mathematics and mechanics by A. M. Lyapunov and A. N. Krylov, and in mathematical analysis, in particular, by that outstanding mathematician, V. A. Steklov. Theoretical discoveries of tremendous practical importance in the field of aerodynamics were made by N. E. Zhukovsky, S. A. Chaplygin, and K. E. Tsiolkovsky. P. N. Lebedev's work on ultra-short radio waves and on light pressure made him famous as one of the world's finest experi-mentators. The attainments of the older generation of physicists were paralleled by a number of important works that brought prominence to scientists of the rising generation. Such were P. P. Lazarev, who initiated modern physical research into photochemical processes; D. S. Rozhdestvensky, who elaborated an ingenious method, since established as a classic, for the quantitative determination of anomalous dispersion of metal vapor;

A. F. Joffe, who became widely known for his experiments in the field of photoelectricity and the physics of crystals. In the natural sciences, Russia could boast such men as K. A. Timiryazev, then engaged in his immortal research into vegetable photosynthesis; the famous selectionist and geneticist I. V. Michurin, and the Darwinian zoologists M. A. Menzbir and A. N. Severtsov. In the field of geology and mineralogy, the Russian natural sciences were represented by the famous crystallographer E. F. Fyodorov; by the "father of Russian geology," A. P. Karpinsky, and by the founders of geochemistry, A. E. Fersman and V. I. Vernadsky.

At the same time, the Russian scientists were constantly haunted by a sense of futility, of unwantedness, of divorcement from their native soil — the inevitable consequence of old Russia's social order and of the tsarist government's fatuous disregard of science. On January 13, 1905 (Old Style), just a few days after the tragedy of Bloody Sunday, P,, N. Lebedev wrote from Moscow to his old friend, Academy member B. B. Golitsyn, in St. Petersburg: "All my efforts to cultivate science in my beloved motherland strike me as a sort of insipid and useless waste of time. I feel that as a scientist I am perishing, beyond salvation. The life that surrounds me is an interminable, stupefying nightmare, hopeless despair. If there is any talk in the Academy about the advance of science in Russia, tell them, from an unhappy Moscow professor, that there is nothing of the kind — no advance, no science, nothing."

These lines were written six years before the police invasion of Moscow University. Coming as they do from the pen of a famous Russian physicist, they .speak eloquently of the tragic gulf that separated science and scientists, on the one hand, and the state, on the other, in pre-revolutionary times.

The extensive publicly supported scientific activities briefly described above, activities always, openly or covertly, directed against the tsarist government, yielded rich fruit when the barriers isolating science from the people were broken down by the proletarian revolution.

*   *   *

The storm that finally broke over Russia in October 1917 put an end to the past and introduced an altogether new life.

To the victorious revolution, science was indispensable for the consolidation of victory and the advance of the new state. The educational level of the masses had to be raised, the semi-literacy handed down from the past to be wiped out, without delay. And the first requirement for this end was schools, schools of every type — elementary, secondary, higher. Teachers and professors had to be trained in tremendous numbers to fill the gap in pedagogical forces. Again, it was urgently necessary that productivity be raised as rapidly as possible, that new productive forces be sought out and brought into use, that the means of production be improved and multiplied. For this, too, science was needed — science untrammeled, research workers, engineers. Scientific institutes and laboratories had to be built and equipped to meet the new requirements.

Barely established, the young Soviet Republic found itself encircled by a hostile ring of imperialist powers. Forces of intervention invaded Soviet territory and came to the support of the White Guard armies. Defense became the order of the day. And in defense as well science was indispensable, improving and modernizing military equipment.

In technology and the natural sciences, much of the heritage coming down from pre-revolutionary learning and of the attainments of science in other countries could be put to work to assist in the titanic labor of building the socialist state. In the social

19

sciences, however, the situation was very different.

The development of Russian science, and particularly of the materialist natural sciences, was strongly and beneficently influenced by the works and ideas of the great Russian revolutionary democrats — Belinsky, Herzen, Chernyshevsky, Dobrolyubov, and the pioneer of Marxism in our country — Plekhanov. It was Russia that gave the world the genius of Lenin and Stalin, those great men of science, who further developed the immortal teachings of Marx and Engels and blazed new trails in human history. In the universities and in the Academy, however, the content and structure of the social sciences were molded to a very great extent by the class views of the capitalist society which had produced them. The idealist philosophy, which, in one form or another, reigned supreme in the Russian universities before the revolution, was a philosophy of a clearly expressed class nature, a fallacious philosophy, hostile to revolutionary consciousness. It was in line with an entirely different philosophy, the dialectical and historical materialism of Marx, Engels, Lenin and Stalin, that the socialist revolution grew and developed. Hence, there was to follow a protracted battle of ideas. Philosophy, history, economics, law all the social sciences — called for swift and radical revision. The science of history as developed in old Russia unquestionably had substantial achievements to its credit so far as the accumulation of factual material was concerned. But the treatment of this material, its interpretation, the theory of historical development applied to it reflected the interests of the nobility and the bourgeoisie. Much that was contained in this science called for revision and renewal. Still more evident was this necessity for fundamental revision in the field of economics and law, which had to be virtually created anew in the conditions prevailing in the world's first classless, socialist state. A new, Marxist-Leninist science was needed here.

Thus, while the natural sciences and technology, for the beginning, at any rate, called only for serious attention, encouragement and material support, the social sciences required a fundamental revision of all that had been handed down from the past.

From the very first days of Soviet rule, it became clear to scientists in Russia that an entirely new phase had begun in scientific development. In the Soviet, socialist state scientific effort was no longer dependent on private initiative or "philanthropical" support. It became ever more clearly an affair of state, a matter of the first importance, an object of particular solicitude to the Soviet government and the Communist Party.

The great majority of Russia's scientists, old and young, realized quickly enough the significance of the great change that had taken place, and appreciated the prospects opening up for science. These scientists soon set to work in the new conditions.

One of the first manifestations of this altogether new turn in the history of science was the rapid organization of large numbers of research institutes. To the tsarist government, as we have already noted, scientific institutions were little more than centers for supplying the necessary teachers, professors, and engineers. Scientific research, the quest of the new, the blazing of trails in science were regarded as the scientist's private affair, not as an essential part of his profession. As a result of this, research, not only in the universities and colleges, but even in the Academy, was generally episodic, confined to scattered individual efforts — what might today be called: "outside the plan." Neither the proper equipment nor the necessary auxiliary personnel were available for scientific work. Such were Mendeleyev's working conditions, Timiryazev's, Pavlov's. Such were the conditions for most of the Russian scientists. P. N. Lebedev was perhaps the first, and almost the only scientist in old Russia who succeeded, despite these difficulties, in founding his own scientific school. He organized a laboratory of considerable size, in which a large number of students and young scientists worked under his guidance. Even in this case, however, the Moscow University Institute of Physics could find no better accommodation for the remarkable works of Lebedev and his pupils than a basement room; and after the events at Moscow University the laboratory was broken up, and had to be reorganized on private funds, in a private apartment.

The changed status of science under Soviet rule was immediately apparent in the radically mew attitude of the .government towards scientific research and its role in the life of the state. From the very beginning, the Party and the government provided extensive and concrete means for the organization of large research institutes, to be independent both of the universities and of the Academy. The first few years of Soviet rule brought into being an entirely new network of scientific institutions — specialized research institutes. The first of these to be organized was the Institute of Physics in Moscow, headed by P. P. Lazarev, based on the privately endowed Institute of Physics that had been established to carry on the work of Lebedev's laboratory. Then, in Petrograd, came the Physico-Technical Institute, headed by A. F. Joffe, and the State Optical Institute, headed by D. S. Rozhdestvensky. Soon afterwards, the Central Aerohydrodynamics Institute (TSAGI) was set up in Moscow, —with N. E. Zhukovsky and S. A. Chaplygin as its leading spirits. Tli en came the All-Union Electro-Technical Institute, in Moscow, headed by K. A. Krug. Big research institutes soon began to appear in other fields of science as well — chemistry, biology, geology. All these institutes were organized and equipped with amazing speed. The Soviet government's appropriations for science were far beyond anything Russia had seen before.

A distinctive feature of the new institutes was the close contact they maintained, through the people's commissariats and the plants and factories, with the targets and problems of the national economy. They became an important link between science and the needs of the state. Thus, the Central Aerohydrodynamics Institute laid the groundwork for the huge Soviet aviation industry. The State Optical Institute rendered great assistance in the development of the optical industry and the improvement of its output. The AH-Union Electro-Technical Institute paved the way for a national electrical industry. The work of the Karpov Chemical Institute, in Moscow, promoted the development and consolidation of various branches of the chemical industry. The Institute of Plant Breeding worked on the problems of intensified farming. Nor did these practical activities crowd out theoretical work, Which also brought splendid results. The institutes became an excellent training school for new scientific personnel recruited from the student and factory youth. And so, around the old Academy of Sciences — former monopolist in the field of purely research institutions — there grew up a large and varied network of scientific centers of an entirely new type, engaged in vigorous research work.

But the Academy, too, had been changing fundamentally since the establishment of Soviet rule. Early in 1918, the Academy of Sciences addressed itself to the Soviet government, expressing its readiness to participate in economic, statistical, and cartographical activities and to undertake research in the field of mineral resources, power production, irrigation, and agriculture. Accepting this offer, the Council of People's Commissars adopted a decision providing for the necessary assistance to the Academy. A rough plan for the Academy, in Lenin's hand, exists to this day — a remarkable document, proposing that the Academy of Sciences be called upon:

"To set up a number of committees of experts to draw up,

as quickly as possible, a plan for the reorganization of industry and economic revival in Russia.

"This plan should provide for :

"Rational distribution of industry in Russia from the point of view of proximity of raw materials and minimum waste of labor in the passage of the raw materials from the initial processing to all subsequent stages of manufacture, up to the final product.

"Rational — from the point of view of the most up-to-date,, largest-scale industry, and particularly trusts — amalgamation and concentration of production in a small number of big enterprises.

"Securing the present Russian Soviet Republic (without the Ukraine and without the German-occupied regions) the maximum possibility to supply itself independently with nil the most important types of raw materials and industry.

"Particular attention to the electrification of industry and transport and the application of electricity in agriculture. The utilization of the poorer grades of fuel (peat, low grades of coal) to produce electric power with the minimum expenditure on the extraction and transportation of fuel.

"Water power and wind-driven motors generally and as applied to agriculture."

And the Academy attacked the problems set before it, to

the extent of its abilities at that time. Ethnographical tables and charts were drawn up, and special committees set to work on the simplification of Russian spelling and the reform of the calendar. Despite the difficulties caused by civil war, the Academy undertook a thoroughgoing investigation of the Kursk magnetic anomaly, which led to the discovery of enormous deposits of iron ore, hitherto unknown. This work was directed by Academy member P. P. Lazarev. who had the active assistance of Academy member A. N. Krylov and of many geologists and geophysicists. A geological survey of the Kola Peninsula, under Academy member A. E. Fersman, brought to light large apatite deposits.

The tiny laboratories of individual Academy members, the Academy's departments and museums, underwent a rapid transformation, growing up into scientific institutes, plentifully staffed and supplied with new equipment, facing entirely new tasks. Thus, the old physics laboratory became the Institute of Physics and Mathematics, headed, at first, by Academy member V. A. Steklov. Later, it was reorganized into three separate institutes: the Lebedev Institute of Physics, the Steklov Institute of Mathematics, and the Seismological Institute. At the proposal of Academy member N. S. Kurnakov, an Institute of Physical and Chemical Analysis was set up. Professor L. A. Chugayev became the head of the new Platinum Institute, which, besides its specific study of platinum, engages in a profound investigation of complex chemical compounds. Academy member V. I. Vernadsky became the head of the Radium Institute. I. P. Pavlov's physiological laboratory grew into a big Physiological Institute. To further the study of language and mentality, Academy member N. Y. Marr organized within the Academy of Sciences the Institute of Language and Mentality, which has elaborated a new theory initiated by its founder. Thus the Academy, formerly at the head of little more than deserted museums, archives, and libraries, was

25

transformed into a broad association of research institutes, populous and active, pursuing clearly defined aims in clearly defined fields.

The Soviet state devoted great attention to the schooling system.

"We can build Communism," said Lenin, in his speech at the Third All-Russian Congress of the Young Communist League, on October 2, 1920, "only from the sum of knowledge, organizations and institutions, only with the stock of human forces and means that were bequeathed to us by the old society. Only by radically recasting the teaching, organization and training of the youth can we ensure that the result of the efforts of the younger generation will be the creation of a society that will be unlike the old society, i.e., a communist society."

New colleges and universities sprang up in all parts of the country. In some cases, these were even organized too hastily, and insufficiently staffed. Trained people were needed, desperately needed, and people were trained by every possible means, including the organization of short-term courses to supplement the ordinary higher schools. The need for teaching personnel in the higher schools was aggravated by the fact that many professors and lecturers had shifted their scientific activities from the old higher schools to the new research institutes in the various branches. But the problem of personnel, desperate as it may at first have seemed, was solved. Within the first decade after the revolution the number of scientific workers — i.e., of persons actively and successfully engaged in research increased some ten times over, at the very least, as compared with, pre-revolutionary limes. It may be said that this eager advance of science in the first years of Soviet rule took as its motto the following statement, made by Lenin at a meeting held in Petrograd on March 13, 1919:

26

"We must take the entire culture that, capitalism left behind and build Socialism with it. We must take all its science and technique, all its knowledge and art. Otherwise we shall not be able to build communist society. And this science, technique, and art are in the hands and in the heads of the experts."

The first Soviet years, the years of civil war and of struggle against the intervention, were a period in themselves for science. At this time the Soviet Union was cut off from the outer world by a hostile capitalist blockade. Hence, no new scientific literature or equipment came into the country, and in this sense, for several years, Soviet science was completely isolated, left to make its way alone. Yet, even in these difficult and exceptional conditions, scientific work not only continued, but developed far more extensively than before the revolution. This period produced a number of works of great moment. It was at this time, for example, that Academy member V. A. Steklov published his research in mathematical physics, and the theoretical physicist A. A. Friedman his important amendments to the general theory of relativity. In Leningrad, the study of atomic structure was under taken on a broad scale. D. S. Rozhdestvensky arrived at very interesting conclusions concerning what is called the fine structure of spectrum lines. When communication with foreign countries was renewed, it transpired that the Soviet physicists, working entirely independently, had in many respects advanced the study of atomic structure. We have already mentioned the thorough experimental investigation of the Kursk magnetic anomaly. In wealth of material, and in the quality of this material, the Kursk investigation has served as a model for many succeeding works of this nature. The chemistry of complex compounds was greatly advanced by the work of Chugayev and his school. It was in this period, too, on the.

instructions of V. I. Lenin and J. V. Stalin, that a group of Soviet electrical engineers, technologists, economists, hydrotechnicians, and construction engineers elaborated the renowned GOELRO plan for the electrification of Russia, of which Stalin wrote :

> "A masterly draft of a really unified, a really state economic plan. ... The only Marxist attempt in our day to build up for the Soviet superstructure of economically backward Russia a really practical technical and production foundation — the only foundation feasible in the present conditions."

This intensive and absorbing work stimulated the growth of new scientific cadres. Young people came pouring into the classrooms and laboratories of the newly organized institutes. They helped to equip the institutes, and, simultaneously with their studies, contributed to the advancement of science. The publication of scientific literature, both original and translated, grew to unprecedented dimensions. Branches of industry which technical backwardness, prior to the revolution, had kept in an embryonic state, now rapidly developed and grew. Such, for example, were the electrical and optical industries. Suffice it to say that before the revolution the country was unable to produce incandescent electric bulbs. Tentative efforts undertaken in this direction shortly before the revolution were a complete failure. In the new conditions, this problem was soon solved, and it was only a few years before the country was fully supplied with bulbs of domestic manufacture. Again, Russia before the revolution had almost no experts in the field of optical instruments. There were but a few small workshops producing such instruments, and even these were mere branches of foreign firms. In the new conditions, experts were soon trained, and the technological difficulties connected with the manufacture of optical glass overcome. The Soviet optical industry began to stand on its own legs. After about ten years, our country no longer had to buy optical glass abroad, though many other countries, both in Europe and in the Americas, still depended on such imports. The chemical industry, too, developed rapidly.

In 1925 the Russian Academy of Sciences celebrated its bicentenary. In connection with this anniversary, it was renamed, now becoming the All-Union Academy of Sciences. The bicentenary festivities, a landmark in scientific life both in the Soviet Union and internationally, were attended by representatives from many lands. A high point of the celebration was the speech delivered by M. I. Kalinin. Congratulating the Academy on behalf of the Soviet government, Kalinin declared that "socialist society, more than any other form of society, urgently requires the broad development of both the abstract and the applied sciences; and it is the first form of society to create for scientific thought and labors genuine freedom and fruitful contact with the masses." On acquaintance with academic institutions, and also with the new, independent institutes which had sprung up since the establishment of Soviet power, it became clear to the assembled scientists, Soviet and foreign, that in a few short years Russia's old science, so limited despite its merits, had grown up into a big new science, steadily and rapidly advancing a science now not only in scope, but in its very nature. Science had become the property of the people, accessible to ail who had the desire and the ability to undertake such work. Every year increased the proportion of students and scientists coming from the working class and the peasantry, both in the universities and colleges and in the research institutes. From the very outset, popularization of science was undertaken on a wide scale. Besides the extensive publication of popular scientific literature and the organization of lectures, this included such methods as the despatch of railroad cars, fitted up with graphic displays aimed to popularize various branches of science, to all parts of the country. Again, with the advance of radio, the Soviet government received still another powerful instrument of political and scientific propaganda.

Another peculiarity of Soviet science was its "practicism" — its contact with the national economy, and its work on problems set by government departments and by branches of industry. Science was definitely entering into the service of the socialist state.

A new method, applied more and more frequently, was that of collective work, in which the solution of a problem would be undertaken, not by one individual, but by a group of scientists, usually headed by a prominent specialist in the field. This method of work made it possible to undertake intricate and laborious research which had formerly seemed impossible.

At the same time Soviet science produced many weighty and outstanding individual works. Academy member A. F. Joffe launched a new approach to that major problem of physics and technology, the strength of crystals, which he attacked by original and ingenious methods of experiment. The young physicist D. V. Skobeltsin (since elected to the Academy) worked out a new and extraordinarily productive method for the study of elementary charged particles in the Wilson chamber in magnetic fields— By this method, Skobeltsin produced the world's first clear and convincing proof of the existence of cosmic rays, and discovered a number of phenomena, hitherto unknown, connected with these rays. It was in this period that Academy member S. V. Lebedev began his efforts to produce synthetic rubber, which were, to culminate so successfully. In this period, too, I. P. Pavlov and his pupils continued their remarkable study of conditioned reflexes, and N. Y. Man's new theory of language was greatly advanced. Soviet science was gaining strength and rallying cadres. It could now go on to the solution, of new problems of great importance to the state.

*   *   *

The first decade of Soviet rule began with a period in which the revolution, politically victorious, had to be defended with arms in hand against its enemies, within and without. Then came a period of reconstruction of the national economy, which had been reduced to ruin as a result of the first world war and the civil war. In many instances science, now rapidly progressing, was able to afford concrete assistance in the reconstruction work in industry, transport, and agriculture, not to speak of its contribution to the people's cultural advancement. Still, the development of science in this first decade was uneven, unsystematic, sometimes entirely spontaneous. Science in the first Soviet decade was not planned and integrated as it is today.

At length, however, the reconstruction work approached completion. The national economy neared its pre-war level, i.e. the state of Russia as of 1913. This fell far short even of the most moderate desideratum. There could be no question, of course, of stopping here. The country was confronted with the urgent task of building up an economy of an entirely new scope, of an entirely new type — the economy of the socialist state. In December 1925, at the fourteenth congress of the Communist Party, Stalin advanced the slogan of industrialization, formulating the work to be done in the next few years in the statement:

> "The conversion of our country from an agrarian into an industrial country able to produce the machinery it needs by its own efforts — that is the essence, the basis of our general line."

Now began a period of intense labor by the whole people, workers, peasants, intellectuals, unswervingly directed towards socialist industrialization.

First of all, the country had to build up a heavy industry — a task of no mean size and difficulty. If the U.S.S.R. was to be independent of the capitalist world, big machine-building, machine-tool, iron and steel, and electrotechnical plants had to be built without delay, and new sources of power to be found and put into use as quickly as possible. Coal and oil output had to be increased, and huge dams and other hydrotechnical projects built. Metals, ferrous and nonferrous, were needed in tremendous quantities.

Down through the ages, Russia had always been an agrarian land. Side, by side with the development of industry, therefore, there arose the problem of stimulating agricultural productivity. And the fifteenth Party congress, in December 1927, adopted a decision, proposed by Stalin, calling for the energetic development of collective agriculture. There followed an extraordinary increase in the demand for agricultural machinery, particularly tractors. Huge tractor plants had to be built to satisfy this need.

The decisions on the industrialization of the country and the collectivization of agriculture were the precursors of the live-year plans. A further decision adopted by the fifteenth Party congress, also on Stalin's suggestion, called upon the State Planning Commission to draft the first five-year plan for the national economy.

In April 1929 the first Stalin five-year plan was approved and adopted.

"The fundamental task of the five-year plan," Stalin pointed out in later years, "was to create such an industry in our country as would be able to re-equip and reorganize, not only the whole of industry, but also transport and agriculture — on the basis of Socialism."

The plan was grandly conceived. And it was carried out, not in five years, but in four. It was followed by a second five-year plan, and a third. With their accomplishment, Socialism was built in our country, and classless society established.

The system of extensive national economic plans calculated for several years ahead brought a new era in Soviet science and technology. The state called upon scientists and engineers for the urgent solution of big new problems, of signal importance in the fulfilment of the five-year plans. And, inevitably, these calls of the state brought the planning principle into science as well.

To Soviet scientists, the past decades have made the idea of planning in science a natural and accustomed concept, an essential attribute to their work. Abroad, however, this idea continues to be a subject of heated debate, arousing no little ideological opposition. An important factor conditioning this refusal to accept and understand the idea of planning in science is to be sought in the individualistic traits of capitalist society, in the cult of private property. Every advance in science, every new scientific idea and invention, in capitalist society, is regarded as an item of private property. The state has no jurisdiction over it, and its development, consequently, cannot be planned. There is no possibility, of course, of planning out "unexpected" scientific results and discoveries; but all true science must contain a very large proportion of well-founded anticipation and prevision. In the seventeenth and eighteenth centuries, for example, Newton's physics might have served as a basis for predicting, and hence planning, the development of physics for a long time ahead. Our contemporary knowledge of the structure of the atom nucleus allows us to plan out for many years to come, with a large degree of confidence, much of the theoretical and experimental work to be done in this field. Contemporary organic chemistry is so constructed that we can see clearly into the future, selecting the most expedient and interesting directions for development in both the practical and the theoretical sphere. In aeroplane construction, empirical formulae have actually been worked out indicating the increase in power of aeroplane motors with the passage of time. Planning is fully `warranted, even essential, in a number of branches of biology, as for example in animal and plant selection, when the question arises as to the desirability of producing one or another breed or type.

And the complete dedication of our science to the service of the people and the state has made planning in science an absolute necessity. That is one of the chief distinguishing features of science! in the socialist: state. Such planning includes not only scope — institutions, personnel, equipment — but also content, i.e., the themes of scientific research.

The plan of scientific development in a socialist state must, of course, link up with the state economic plan; but it should not be forgotten that the prospects opened up by the constant growth of science will often considerably exceed the prospects outlined in economic planning. Science has its own peculiar logic of development, a logic which it is essential to take into account. Science must always work ahead, accumulating reserves for the future; only then will it be working in its natural element.

The definite transition to the planning principle was the chief distinguishing feature of Soviet science during tin? second period of its development, approximately coinciding with the second decade of Soviet rule. Another important feature of this period was the gradual decentralization of science, the appearance of new hubs of scientific activity. It was at this time that the Academy of Sciences of the U.S.S.R. set up its first branches : the

Far-Eastern branch, in Vladivostok; the Urals branch, in Sverdlovsk; the Georgian branch, in Tbilisi; the Armenian branch, in Erevan; the Azerbaijanian branch, in Baku, and the Kazakh branch, in Alma-Ata. These were designed to promote the development of scientific research in various directions, as determined by local conditions and requirements; they provided a research set-up to complement the existing local universities and colleges. In `the course of time, the branches fully justified their existence. They concentrated the work of local scientists, trained new forces from among the local population, and soon began to produce important results, both theoretical and practical. In later years a number of these branches developed to a point at which they could be reorganized as independent Academies.

The decentralization of science affected not only the Academy, but also the network of specialized research institutes. Several big institutes of this type came into being in different parts of the country under the five-year plans. Special mention should be made of the physico-technical institutes in Kharkov, Dniepropetrovsk, Sverdlovsk and Tomsk, the organization and staffing of which were greatly facilitated by preparatory work conducted by the Leningrad Physico-Technical Institute, under Academy member A. F. Joffe. These four institutes became important scientific centers, and in a brief space of time produced results of considerable significance. Big agricultural institutes were also set up in various parts of the country, in Omsk and Odessa, for example, as well as a number of other institutes in various branches and specialties.

A third distinguishing feature of Soviet science in the period of the first Stalin five-year plans was the great increase in the number of universities and colleges and in the student body. Pre-revolutionary Russia could boast only 91 universities and colleges, with a student body of some 112,000 (figures for

1914-1915). When the first five-year plan was launched, in 1928-1929, the student body numbered about 177,000. By the beginning of the second five-year plan, in 1933-1934, this figure bad jumped to 504,000. By the beginning of the third five-year plan (1937-1938) it had risen to approximately 603,000. And on the eve of the Patriotic War, in 1941, there were some 800 universities and colleges in the Soviet Union, with a student body of 667,000. Thus, in the course of three Stalin five-year plans, the Soviet Union's college and university student body multiplied almost four times over. It should be added that in 1941 there were also some 12,000 post-graduate students, i.e., future scientists and research workers, enrolled in the different universities, colleges and scientific research organizations.

Simultaneously with the introduction and consolidation of planning, with the process of decentralization, and with the rapid increase in scientific personnel, the country's scientific research network underwent a process of differentiation and clarification of functions. Questions of scientific principle were now concentrated chiefly in the Academies, central, republican, and specialized. The colleges and universities occupied an intermediate position. Dedicated primarily to the work of training scientists, teachers, engineers, they at the same time engaged in research activities, both in the sphere of general theory and along practical lines in the various specialties. Principally, however, the practical, technical solution of the problems brought forward in every field by the development of the national economy was concentrated in the big specialized institutes and in the factory laboratories, which maintain direct contact with industry.

In the summer of 1934, in accordance with a decision of the Council of People's Commissars dated April 25 of the same year, the Academy of Sciences of the U.S.S.R. moved from Leningrad to Moscow. This change in location, after more than two hundred years of work on the banks of the Neva, was fully in line with the momentous changes which had come about in the very nature of the Academy's work. Actually, the Academy now headed a national, state scientific research network, and its activities were closely linked up with the concrete problems facing the Soviet state. The new charter of the Academy of Sciences, confirmed by the government on November 23, 1935, describes the Academy of Sciences of the U.S.S.R. as "the highest scientific institution of the U.S.S.R., uniting in its ranks the country's most outstanding scientists." The fundamental purpose of the Academy is defined in the new charter as : "...universal assistance in the general advance of scientific theory and of the applied sciences in the U.S.S.R., and the study and further development of scientific achievements attained in other countries. The Academy of Sciences regards as its basic task the systematic application of scientific achievements to promote the building of the new, socialist, classless society."

Particularly striking, in connection with these changes in the nature of the Academy's work since the establishment of Soviet rule, with the close contact established between this work and the requirements and ideology of Socialism and the Soviet state, is the fundamental revision undergone by the social sciences as represented in the Academy, both in content and in trend. In 1936, in view of such revision, the research institutes of the Communist Academy became a part of the Academy of Sciences of the U.S.S.R. The State Academy of the History of the Material Aspects of Civilization has also been taken over by the Academy of Sciences.

As we have noted above, the five-year plans brought a continued growth in the work of the specialized institutes, and a steady increase in their number. Gradually, as the result of protracted and strenuous effort in many fields, a situation was brought about which might be called, figuratively speaking, an uninterrupted front in science and technology. Pre-revolutionary Russian science had every right to pride itself on its individual great names, and on the individual fields of work in which it attained remarkable successes. At the same time, however, there were many branches of science and technology in which old Russia, at times, could not boast a single specialist. In such cases, aid would have to be requested abroad. Thus, .Russian technology was in many respects dependent on favors — often anything but disinterested — from other countries. The establishment of an uninterrupted front in science, the training of young scientists and engineers specializing in every conceivable field, is a labor of extreme difficulty, which only a very few countries have accomplished. The establishment of such an uninterrupted front in the period of the first five-year plans, and the appearance of specialists in almost every field of practical importance, was one of the most outstanding achievements resulting from the planning principle in the development of Soviet science and technology. Its attainment required protracted and highly differentiated training; it required independent effort on the part of scientists and engineers; again, it required close and constant contact with industry, the joint exertions of science and industry to overcome obstacles and difficulties.

An important index of the vigorous growth of Soviet science during the five-year plans was the tremendous progress in the sphere of specialized publications. As yet, unfortunately, there has been no bibliographical survey of our scientific literature and its development since the establishment of Soviet power. Still, a general knowledge of the quality of this literature, and a comparison with developments in other countries during the same period, permit the confident assertion that our attainments in this field are very considerable indeed. Just one example: In tsarist Russia there was only one periodical devoted to original scientific papers in the sphere of physics. This periodical had no more than

200 subscribers. At present, there are five big periodicals devoted to physics, with a circulation of some 5,000 each. The same is true, often in even more astounding measure of other fields of science. During the pre-war five-year plans, public activity in science was very intense. This period was marked by an endless succession of congresses, conferences, and meetings, in many cases initiated and organized by the Academy of Sciences of the U.S.S.R. In 1940, for example, the Academy of Sciences organized 70 conferences attended by representatives of different scientific institutes and industrial plants. The Academy held special sessions in Sverdlovsk and Novosibirsk on the problem of the Urals and Kuzbas. Another session, held in Leningrad, was devoted to the problem of the Volga and the Caspian. The number of expeditions sent out 'by the Academy of Sciences and various special organizations increased yearly in every field: flora, fauna, geology, geography, ethnography, archaeology. Again, there were many complex expeditions, as for example the one to Mt. Elbrus, which included representatives of the most varied branches of science, from specialists in the study of cosmic rays to physiologists and physicians.

The tremendous advance of science and technology under the five-year plans, and the uninterrupted front of science which took shape in this period, make it practically impossible to describe, indeed, even to enumerate, in such short space, even the main attainments of Soviet science, as embodied, on the one hand, in mountains of books, periodicals, patents, and copyrights, and, on the other, in concrete form, in machines, factories, foods, and goods. Hence, the present resume must be limited to a very brief and superficial review of a few particularly outstanding works.

Russian mathematics had occupied a leading position in world science ever since the beginning of the nineteenth century; but never had it attained such scope, such depth, such variety as in the period we are now examining. We must note especially the original works of our mathematicians, and in particular those of Academy member I. M. Vinogradov in the theory of numbers: the development of a new analytical method, and the solution of several extremely difficult problems in this field. Academy members S. N. Bernstein and A. N. Kolmogorov, and corresponding Academy member A. Y. Khinchin, produced works in the theory of probabilities of great importance not only to mathematics, but also to physics, various branches of statistics, technology, and the military sciences. Important progress, of great practical significance, was scored in the theory of differential equations. Of the many brilliant works in this field, we may note those of Academy members I. G. Petrovsky, S. L. Sobelev; and V. I. Smirnov. New and original work was done in geometrical topology by P. S. Alexandrov, a corresponding member of the Academy.

Before the revolution, Russian physics had developed in but few directions. In the new conditions, however, it quickly expanded along a broad and inclusive front. In the U.S.S.R. today the physical sciences are represented by numerous specialists, the authors of important developments both in general theory and in practical technology. Among the most outstanding of these developments in the period under consideration, first mention must be accorded to the remarkable discovery by Academy members L. I. Mandelstam and G. S. Landsberg of a new form of diffraction of light, which has received the name of combination diffraction. Simultaneously with the Soviet scientists (1928), this phenomenon was discovered also by the Indian physicist Raman, in Calcutta. It laid the foundation for a new and very extensive field of science, interesting both physicists and chemists, and opened new possibilities in the study of molecular structure. Soviet researchers also attained momentous results in the study of physical

40

phenomena occurring at temperatures approximating to absolute zero. Thus, Academy member P. L. Kapitsa discovered a new and remarkable property of liquid helium, which has been called "superfluidity." The theoretical elucidation of this astonishing phenomenon was worked out by Academy member L. D. Landau; and the most subtle conclusions to which this theory leads (two sounds in liquid helium) were confirmed experimentally by the young physicist V. P. Peshkov.

Soviet physicists and mathematicians made basic contributions to the study of non-linear oscillations, i.e., oscillations mathematically expressed by non-linear differential equations. The works of Academy members L. I. Mandelstam, N. D. Papalexy, A. A. Andronov, and N. M. Krylov, and of corresponding member N. N. Bogolyubov, led to most important conclusions, both theoretical and practical, in radio and mechanics.

Academy member A. F. Joffe completed a number of important systematic investigations in the physics of semi-conductors, which opened up new prospects in the field of electrotechnical materials, photoelectricity, and the like. After a profound and detailed investigation of interference phenomena, Academy member V. P. Linnik succeeded in working out a large number of ingenious interference instruments, based on new principles, which offer important possibilities in the study of the quality of surfaces, of the precision of mechanical parts an the construction of astronomical instruments, and the like.

Soviet chemistry also developed and expanded under the Stalin five-year plans, and produced a number of works of the greatest moment, both theoretical and practical. Thus, the works of Academy members A. E. Favorsky and S. V. Lebedev paved the way for the establishment of the synthetic rubber industry in the U.S.S.R. The investigations of Academy member A. N. Nesmeyanov threw a new light on the important field of organometallic compounds. Academy members N. D. Zelinsky and A. A. Balandin completed work of some importance, both theoretical and practical, in the field of catalysis. A number of new and important trends developed in Soviet physical chemistry. In the study of surface-active substances, note should be taken of the numerous systematic investigations of Academy members A. N. Frumkin and P. A. Rebinder. Academy member N. N. Semyonov greatly advanced the study, both theoretical .and experimental, of chain reactions and their kinetics. Academy member A. N. Terenin did important experimental work on photochemical reactions. He is the discoverer of the photodissociation of diatomic molecules, and has secured very promising results in the field of complex organic compounds.

The collective style of work which has become a feature of Soviet science generally was particularly notable in the huge geological investigations conducted under the five-year plans. It was these investigations, seeking and discovering oil, metal ores, and other minerals in various parts of the Soviet Union, that charted Soviet industry's raw material base. The labors of Academy members A. D. Arkhangelsky, I. M. Gubkin, S. S. Smirnov, P. I. Stepanov, A. E. Fersman, and V. A. Obruchev, and their numerous pupils and followers, made possible the solution of many problems of the first importance involved in the fulfillment of the five-year plans.

Particularly important among the numerous geographical expeditions and investigations undertaken in this period was the exploration and conquest of the Arctic, marked by such outstanding events as the voyage of the Chelyuskin, the flight to the North Pole headed by Academy member O. J. Schmidt, and the renowned Papanin camp on the drifting ice.

From the very outset, Soviet biology entered the service of agriculture and medicine. Many outstanding achievements of

Soviet science in the sphere of plant and animal selection, plant breeding, and phytogeography found immediate application in agriculture. Academy member I. P. Pavlov's investigation of high nervous activity, begun prior to the revolution, was greatly expanded in the new conditions, both in his own work and in that of his pupils, and led to many conclusions of great importance in medicine.

Technology, too, numbers countless achievements for this period. The quality and scope of Soviet technology were well expressed in such gigantic power projects as the hydrotechnical stations on the Svir, the Volkhov, and the Dnieper. Industry — iron and steel, machine-building, electrical, chemical — grew up on the basis of Soviet science, of the tremendous experience gained by our scientists and engineers. Powerful radio stations, a reorganized railway system, the Moscow subway, the huge dams and locks of the Moscow-Volga canal — such are a few examples of the new Soviet technology developed under the five-year plans.

The fundamental revision of the social sciences continued under the five-year plans. This period was marked by practical work on the history of the Soviet Union. The history of literature, both of the Russian people and of other Soviet nationalities, was now approached, for the first time, from the point of view of the new, Soviet principles in the study of literature. Orientalogy, in the multi-national Soviet Union, acquired an entirely new trend, being applied to the problem of creating grammars and dictionaries for different Soviet peoples, and of compiling their histories. A new science of law was developed, and intricate problems arising in the new Soviet economics investigated.

All these achievements of Soviet science were facilitated by the exceptional solicitude and attention accorded it by Stalin, and by his views on progressive science, which guided the Academy of Sciences in all its work.

Of tremendous importance in the ideological development of Soviet scientists was Stalin's remarkable work, the History of the C.P.S.U. (B.), Short Course, published in 1938.

In December 1939 the General Assembly of the Academy of Sciences of the U.S.S.R. elected Joseph Vissarionovich Stalin an honorary member of the Academy, for his outstanding services towards the advancement of science and for his continuation and development of the Marxist-Leninist doctrine in every sphere.

This election was a vividly symbolic act, illustrating the transformation which had made the Academy of Sciences of the U.S.S.R. the true staff of progressive Soviet science.

By 1941, on the eve of the Great Patriotic War, the Soviet Union could boast a huge army of scientists, comprising almost a hundred thousand men and women who had dedicated their lives to scientific labors in the countless new institutes, in the Academies, in the colleges and universities, in industry. The Soviet scientists had created a great new scientific literature, and had organized themselves in a scientific front which was destined to aid the military front in the difficult years of war.

*　　　*　　　*

In launching their campaign against the Soviet Union, the fascists miscalculated on a great number of points. One of these was an underestimation of Soviet science.

The war was a test for Soviet science, a test of redoubled severity. On the one hand, science was called upon to solve entirely new and often extremely intricate problems in every conceivable field, set before it urgently by the front, the war industries, and the national economy as a whole. On the other hand, it was compelled to work in unaccustomed conditions,, often involving great hardship.

Many of the Soviet scientists went to the front to defend their country, exchanging books and laboratories for rifles or fighter planes. Many never returned. They gave their lives for their country in the field of battle.

Enemy bombs and shells destroyed the Pulkovo Observatory and the famous conservatories of the Leningrad Botanical Gardens. The astronomical observatory at Simeiz was looted, and destroyed by fire. The German vandals blew up Kiev University and the Byelorussian Academy of Sciences, and plundered the laboratories and libraries of many colleges and universities. The losses of scientific equipment were tremendous.

A large number of scientific institutes were evacuated far to the rear. Here they had to work in unaccustomed conditions, without the proper equipment, instruments, materials, and libraries. In some cases, as for example in Leningrad during the blockade, scientific work was carried on through cold and hunger, punctuated by daily enemy bombardment.

Despite these hardships and privations, Soviet science came through the war with, colors flying. Its response to the wartime requirements was concretely expressed in the form of new and improved types of artillery; in rocket projectiles; in the constant improvement of planes and motors; in the development of new types of armor and of armor-piercing shells to fight the German "Tigers" and "Ferdinands"; in the achievements of Soviet radio; in the faultless service of all types of military optical instruments, and their constant development and improvement; in the splendid organization of the medical service, which saved the lives of hundreds of thousands of wounded soldiers and combated infection and epidemic both at the front and in the rear.

Every new detail of military equipment and materials, every new drug and method of treatment, bore the imprint of scientific thought and labor.

The war industries demanded new and more rapid methods of testing output, new machinery, new materials, new designs; and in almost every case science supplied the need. Agriculture, with almost all the able-bodied men of the countryside off at the front, called for urgent agronomical and agrotechnical assistance, for the development of new methods of work. Here, too, science responded promptly to the call.

The knowledge and experience accumulated in the years preceding the war, the abundance of scientific personnel, and the devoted patriotism of the Soviet scientists helped the country to overcome many difficulties. Nor was Soviet science, in this period, confined entirely to wartime effort — for the front, for industry, for agriculture, for medicine. It also continued its development along fundamental lines. That is clearly to be seen in the long lists of Stalin prizes awarded during the war for outstanding works in science and technology. These remarkable lists record the momentous scientific achievements of industrial personnel, collective farmers, engineers, and eminent scientists engaged in the study of key problems in the different fields of science. Even in the most trying days of the war, scientific thought worked on.

All the leading scientific periodicals in the Soviet Union continued publication throughout the war, and the majority of the universities and colleges continued to function. Early in 1943, at the time of the decisive fighting at Stalingrad, the Soviet scientists marked the three-hundredth anniversary of the birth of Isaac Newton, the great founder of modern physics. Celebrated with the warmest enthusiasm and interest, this holiday of science, at the very height of the war, at the period of its crisis, was a striking demonstration of the strength and vitality of Soviet science.

Like the period of the Stalin five-year plans, the war was a new school for science. It taught' scientists to distinguish even more clearly than before between things of the first and of the second importance, between matters of state and "pure science," so called. The war showed how swiftly and confidently the most difficult of problems can be solved by a collective scientific body inspired by fervent patriotism; it showed what potent scientific forces lay latent in the most far-flung parts of our country.

Early in the war — in the spring of 1942, after the rout of the Hitlerites on the approaches to Moscow — Stalin wrote, in a telegram addressed to the president of the Academy of Sciences:

> "I am confident that the Academy of Sciences, despite difficult wartime conditions, will keep pace with the increased requirements of the country."

In a second telegram to the president of the Academy, Stalin wrote :

> "I hope that the Academy of Sciences will head the movement of innovators in science and industry, will become the center for progressive Soviet science in the struggle which has been launched against the most malignant enemy of our people and of all other freedom-loving peoples — German fascism."

In the Academies, in the specialized institutes, in the colleges and universities, Soviet scientists and engineers strained every effort to justify Stalin's faith in Soviet science, to help the Soviet Army and the Soviet people through the difficult years of war. Despite the difficult and unaccustomed conditions, science kept pace with the country's increased requirements:. Soviet men of science were to be found everywhere — in the air force, the navy, the artillery, the engineers, the railway troops, the hospitals, the war plants, the collective farms. And everywhere they offered help and counsel. Soviet science may claim its share in the victory of the Soviet Army.

The two hundred and twentieth anniversary of the foundation of the Academy of Sciences of the U.S.S.R., celebrated in June 1945, when the red Soviet banner of victory was floating, over the Reichstag in Berlin, came as a red-letter day to Soviet scientists, a day for the summation and review of all that Soviet science had accomplished both in time of war and in the entire post-revolutionary period.

After the victory gained in the Great Patriotic War our motherland, and our science, turned to new tasks. Science was called upon to take an active part in the rehabilitation of towns and villages which the enemy had raided, plundered and destroyed, and in the fulfillment of the post-war Stalin five-year plan of reconstruction and development of the national economy. After the military problems of the recent past, science now turned to the manifold problems of socialist construction. The direction and content of all the work of the Academy of Sciences are defined by the tasks which the great Stalin has, set the Academy.

Soviet science is increasingly well provided for. It enjoys the unfailing support of the Communist Party, the Soviet government, and Stalin personally. An outstanding part in the development of Soviet science has likewise been played by Stalin's closest comrade in arms, Vyacheslav Mikhailovich Molotov.

On November 29, 1946, the Academy of Sciences of the U.S.S.R. elected Vyacheslav Mikhailovich Molotov an honorary member of the Academy, for his eminent contributions to the development of the Marxist-Leninist science of society, the state and international relations, and for his distinguished services in the building and consolidation of the Soviet state.

Heir to everything worthwhile left us by pre-revolutionary Russia in the field of culture, Soviet science has grown up together with the country. It has come through the early Soviet years, the period of intervention and civil war; it has come through the stern school of the Stalin five-year plans, and been tempered by the heroic days of the Great Patriotic War.

Our science met the thirtieth anniversary of the Great October Revolution as a distinctly Soviet science, differing essentially from the science of other countries. Our science is vigorous, extensive, comprehensive. It keeps pace with every need of the Soviet state and the national economy. That is one of its most striking distinctive features, as compared with the science of old Russia.

And what has science given our country in this period? We need only look about us to see its fruits on every hand. Indeed, the Soviet state as a whole, through all the difficulties of existence in capitalist encirclement, is guided and directed along the lines conceived and expounded by the great scientific doctrine of Marx, Engels, Lenin and Stalin.

Soviet science justly prides itself upon the fact that our people have given the world the genius of Lenin and Stalin, supreme exponents of progressive scientific thought, founders of the Soviet state. Lenin and Stalin have enriched Marxism and greatly advanced it, applying it to the new conditions of development of society. They have revealed the laws of this development in the new age, that of imperialism and socialist revolution. They have created the doctrine of the victory of Socialism, and of the building of Socialism in our country on the foundation of the Soviet system, and they have realized this great teaching in practice.

Our science is based on the brilliant scientific works of Lenin and Stalin — in particular, Lenin's *Development of Capitalism in Russia* (1899), *Materialism and Empirio-Criticism* (1908), and *State and Revolution* (1917), and Stalin's *Anarchism or Socialism* (1906-07), *Marxism and the National Question*

(1912-13), *Foundations of Leninism* (1924), and *Dialectical and Historical Materialism* (1938); it is based on Stalin's program for the socialist industrialization of our country and the collectivization of agriculture, on Stalin's military strategy, on Stalin's teachings concerning the state, cadres, Soviet intellectuals, and progressive science, on the great Stalin Constitution. Such is the indestructible foundation of our science, the bulwark of the sole true philosophy, our guide in scientific labor and in the fight for Communism.

Thanks lo the works of Lenin and Stalin, the development of the state, as a social process, is now, for the first time in human history, regulated on the basis of scientific theory. And side by side with this magnificent manifestation of the science of human society in the life of the Soviet state, we observe in every sphere the results of the concrete application of modern science and technology. The simple electric bulb which illumines the expanses of the Soviet land received its present shape as the result of long years of co-operation between science and industry, as the result of manifold assistance from physics, chemistry, the iron and steel industry, the glass industry, and high vacuum technique. Radio, invented half a century ago by A. S. Popov, has developed and expanded as a result of the tireless work of Soviet physicists and engineers. To the Soviet citizen, radio takes the form of super-powerful broadcasting stations and of a tremendous receiving network. It has penetrated to the most distant corners of the country. Radio was a very timely development for the socialist, Soviet land. It has become a powerful means of information and propaganda, a means of uniting the people in labor, struggle, and festivity. Telephones, automobiles of every type, new models of steamers and locomotives, Soviet aircraft, which are undergoing constant modification and improvement — all demonstrate that science and technology have penetrated to the very core of life in our country. Science has in many ways revised the nature of cultivated crops. Thus, it has produced new and improved forms of the cereal crops, adapted to the peculiarities of climate prevailing in different parts of our country. Human lives are preserved by the knowledge and skill of Soviet surgeons and roentgenologists, by Soviet medicines. The clothes we wear, the buildings we live in, the electricity we use — all this is the result of the application and development of our scientific and technical knowledge. In concentrated form this knowledge is expressed in mountains of books published since the establishment of Soviet rule.

The views of Soviet men and women on nature and society have changed fundamentally. They are now based on wholesome, unconquerable dialectical materialism. Guided by dialectical materialism, the Soviet scientist fearlessly combats every attempt to distort science, every manifestation of the fog of idealism which may appear from time to time in the path of scientific development.

But science cannot stop, cannot rest on its laurels. Science, by its very nature, is variable, dynamic, incapable of marking time. And this dynamic force of Soviet science is embodied in its cadres, in its scares of thousands of specialists, including over ten thousand Doctors of Science and some twenty-five thousand Masters of Science. This huge scientific army which has grown up in the years of Soviet rule will create the future of science, will solve the countless new problems facing the Soviet Union in the post-war period.

The second world war brought humanity concrete proof of the tremendous importance of science and technology in our times. The development of science has put into the hands of the human race weapons and natural forces equivalent in power to elemental upheavals. And it is a matter of the most vital moment into whose hands these mighty weapons fall. Science and technology in the hands of insane fascism threatened the fate of humanity. Science and technology in the hands of imperialists who dream of world dominion become a means of enslaving the peoples. Science and technology in the hands of progressive Soviet democracy promote universal prosperity and facilitate the advance to Communism.

In the early eighteenth century Russian science was helped by renowned foreigners — Euler, Bernoulli, and others. Later in the same century, our country evinced its own power. From the depths of the people, from distant Archangel, Kazan, Tobolsk, Ryazan, came the great Russian scientists: Lomonosov, Lobachevsky, Mendeleyev, Pavlov. These men set the world brilliant examples of creative scientific labor. But only since the establishment of Soviet rule have all the country's latent forces been set in motion, have the great, but isolated scientists of the past given place to a huge army of Soviet scientists. It was upon this army that Stalin called, on February 9, 1946, rapidly to overtake and surpass the achievements of other countries. Much has already been done to facilitate the accomplishment of this task. At the same time, the Party and the government are further increasing their assistance to science in the shape of new buildings, equipment, and improved conditions for research work. The Soviet scientists are backed by great experience in the past, and — what is perhaps of even greater importance — they are confronted with a great and absorbing task: to help their country attain, in the shortest possible time, to that most perfect form of social life — Communism. The first three decades of Soviet rule were a period of continuous growth and development of science. The fourth decade must and shall become a period of gigantic scientific achievement. That is our debt to the Soviet people, to our government and Party, to our great leader and teacher, Stalin.

www.ingramcontent.com/pod-product-compliance
Lightning Source LLC
Chambersburg PA
CBHW051250170526
45165CB00004B/1656